AI AWARENESS SERIES

AI in Change Management

Darian Batra

Contents

Introduction

Change is a constant in today's organizations—but the emergence of artificial intelligence is transforming how change itself is understood, designed, and delivered. As AI becomes more integrated into business operations, leadership, and culture, it is reshaping the expectations placed on change practitioners and accelerating the need for new capabilities.

AI in Change Management is written to help change leaders and professionals make sense of this transformation. It explores how AI is not only a technological tool but also a strategic enabler of organizational change. Rather than replacing human insight, AI enhances decision-making, provides deeper diagnostics, supports tailored communication, and enables more adaptive and data-driven approaches to transformation.

This book begins with a foundational look at how AI is impacting the nature of change in organizations and introduces key AI concepts relevant to the change management profession. From there, we explore how AI interacts with each stage of the change lifecycle—from assessment and planning to implementation, communication, and reinforcement.

We also dive into practical applications of AI, including its use in talent and capability assessment, scenario modeling, and change communication. Importantly, we address the ethical implications of using AI in organizational change—highlighting the need for transparency, governance, and human-centered leadership.

Through real-world case studies and future-focused insights, this book helps readers envision how AI will shape the future of change leadership. Whether you're guiding a digital transformation, leading cultural change, or managing strategic shifts, understanding how to leverage AI responsibly and effectively is now a critical capability.

Part of the AI Awareness Series, this book is designed to bridge the knowledge gap between change practitioners and emerging technologies—so you can lead with confidence, clarity, and purpose in the age of AI.

Let's begin.

Chapter 1: Understanding Change in the AI Era

Organizational change in the AI era means more than just adopting new tools. Companies must adapt their structures to integrate AI effectively. Automation now streamlines processes, boosting efficiency. AI also enhances decision-making with data-driven insights, and as AI reshapes roles, it creates opportunities for new job functions and the augmentation of human skills.

Looking back, we can see how technology has consistently driven change. The Industrial Revolution introduced machines that reshaped industries and societies. The internet then transformed how we communicate and conduct business. Today, AI is pushing change even further, with complexity and speed we've never seen before.

AI is accelerating change across all industries at a rapid pace. Organizations are facing constant disruption, which makes adaptability essential for survival. To keep up, businesses must foster a culture of agile learning—always evolving alongside shifting markets and technologies.

Organizations today face intense market competition. AI adoption helps companies innovate, enhance customer experiences, and streamline operations. Meanwhile, digital transformation isn't optional—it's a necessity for staying relevant in a fast-changing market.

Regulations are another key driver of change. Data privacy laws demand strict protection of personal information. At the same time, AI ethics guidelines push organizations to ensure fairness and transparency. Plus, industry-specific compliance standards must be met to maintain trust and avoid penalties.

Internally, organizations are motivated by the need for innovation, efficiency, and cost reduction. AI drives innovation by enabling new products and services. It streamlines operations by automating tasks and improving workflows. And by reducing manual effort, AI can significantly cut costs while supporting growth.

AI and automation help optimize repetitive tasks, boosting efficiency and reducing errors. Predictive analytics allow businesses to anticipate trends and make smarter decisions. These technologies also transform workforce roles, supporting innovative business models and creating new opportunities.

Advanced data analytics provide actionable insights that lead to better business decisions. This improves decision quality by giving organizations accurate, relevant information. With real-time analytics, businesses can make faster decisions and adapt their strategies to a constantly changing environment.

Integrating digital tools ensures seamless workflows, reducing delays and boosting efficiency. Unified platforms enhance collaboration, even across remote teams. And with stronger digital integration, organizations become more agile—ready to respond quickly to new challenges.

Disruption refers to sudden changes that can challenge or even threaten a company's existing business model. While it poses risks, disruption also brings opportunities for innovation and growth. The key is recognizing these changes early and responding swiftly to stay competitive.

Transformation is about proactive change—deliberately shifting company culture to foster innovation and adaptability. By optimizing processes and evolving strategies, organizations can use emerging technologies to secure long-term success and stay ahead of the curve.

We've seen companies both disrupted and transformed by AI. Some struggled as their traditional models became outdated, while others thrived by embracing change and using AI to improve efficiency and seize new opportunities. These real-world examples show how vital effective change management is for successful AI adoption.

Change naturally creates resistance, often due to fear of the unknown. Overcoming this requires clear, honest communication to build trust. Providing training and support helps employees feel confident with new tools. Involving them in the change process further reduces resistance and encourages engagement.

Leadership plays a critical role in guiding change. Visionary leaders provide direction and inspire their teams to embrace transformation. They also motivate teams and lead by example—demonstrating adaptability and flexibility in the face of AI-driven changes.

Finally, organizations need to foster a culture of adaptability and resilience. This means being open to change, encouraging learning, and remaining flexible as technology and markets evolve. Building this kind of culture prepares businesses to thrive in the face of constant transformation.

To wrap up, we've seen that navigating change in the AI era requires understanding its drivers, embracing emerging technologies, recognizing the difference between disruption and transformation, and focusing on human factors like leadership and adaptability. With these strategies, organizations can not only survive but thrive amid AI-driven change.

Chapter 2: Introduction to AI for Change Leaders

Let's start by defining Artificial Intelligence in the context of business change. AI refers to systems that can perform tasks normally requiring human intelligence. This enables businesses to enhance their capabilities and efficiency. AI supports smarter, data-driven decisions, leading to better outcomes. It also optimizes processes by automating routine tasks and improving overall efficiency. Finally, AI can be a catalyst for innovation, unlocking solutions that transform traditional business models.

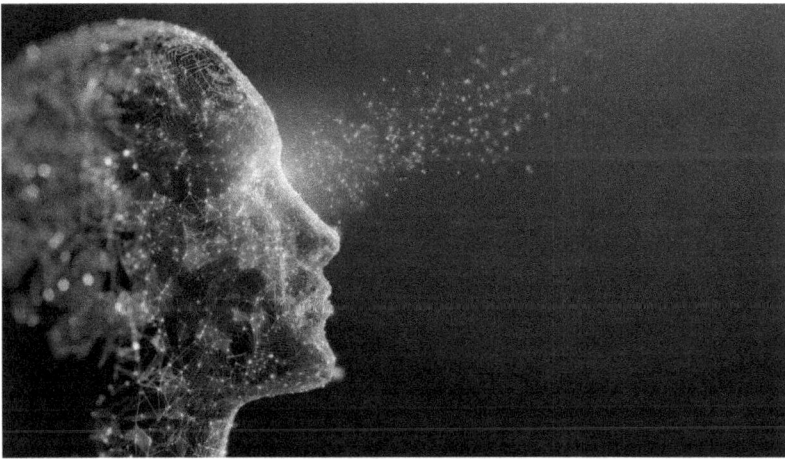

As AI evolves, so does the role of change leaders within organizations. Change leaders are now tasked with integrating AI strategies that align with business goals and future growth. They play a key role in promoting digital literacy, ensuring their teams are equipped to work with AI. It's also vital for leaders to support employees emotionally and professionally through AI-driven change. And when disruption happens—as it often does with AI—effective leaders guide their teams with agility and confidence.

AI adoption brings both exciting opportunities and important challenges. On one hand, AI can dramatically enhance efficiency and spark innovation across industries. However, it also raises concerns about data privacy that organizations must take seriously. Ethical considerations demand strong governance to ensure AI is used fairly and responsibly. Lastly, organizations need to support their workforce as roles shift, ensuring employees adapt and grow alongside AI.

Understanding core AI concepts is crucial for any leader navigating business change. First, algorithms—these are step-by-step procedures AI systems use to solve problems. Neural networks, which mimic the structure of the human brain, help AI recognize patterns and learn. Data training is the process of feeding AI large datasets to improve its performance. And finally, natural language processing allows AI to understand and interact using human language, unlocking many practical applications.

Now, let's distinguish between two types of AI: narrow and general. Narrow AI is designed for specific tasks—like image recognition—and excels within defined boundaries. General AI, still largely theoretical, would perform any intellectual task a human can. In business today, we mainly work with narrow AI, leveraging it to enhance specific functions and processes.

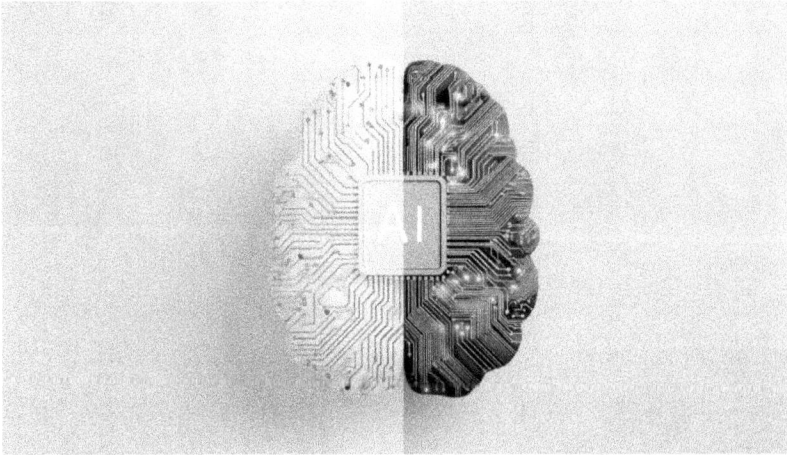

AI needs to align with business objectives to truly drive transformation. It can automate routine tasks, freeing up time and resources for more strategic activities. AI also plays a big role in enhancing customer experience through personalized services and quicker responses. And with predictive analytics, AI helps businesses make informed decisions that accelerate transformation and build a competitive edge.

Let's look at the fundamentals of machine learning—an essential branch of AI. Machine learning allows systems to learn from data without explicit programming. This capability forms the backbone of many business applications we'll cover next.

Machine learning has many practical applications in business decision-making. For example, it helps segment customers for targeted marketing. It plays a vital role in fraud detection, identifying threats in real-time. It also improves supply chain operations by optimizing logistics and reducing costs. And in marketing, machine learning enables highly personalized campaigns based on customer behavior and preferences.

But as powerful as AI is, we must be mindful of ethical considerations. Data bias is a key concern—if AI learns from biased data, it can produce unfair outcomes. Transparency is also critical; stakeholders need to understand how AI decisions are made. And finally, accountability must always rest with human leaders to ensure AI is used ethically and responsibly.

Now let's turn to generative AI—an area that's seen rapid growth recently. Generative AI models create new content based on the data they learn from. One example is GANs—Generative Adversarial Networks—that generate realistic images. Another is transformer models, which excel at language tasks like text generation. These tools enable applications like content creation, image synthesis, and automating design processes.

Generative AI has many valuable applications within organizations. It helps create marketing content quickly, boosting engagement and outreach. In product design, it speeds up prototyping, allowing for rapid innovation. It also enhances datasets through data augmentation, making machine learning models more effective. And by powering AI-driven tools like chatbots, it enhances customer interactions and personalization.

Despite its benefits, generative AI comes with risks and governance challenges. It can be misused to spread misinformation or infringe on intellectual property rights. Ethical considerations demand careful attention to fairness, transparency, and accountability. To manage these risks, organizations need strong governance frameworks and policies that ensure responsible use of generative AI.

Let's explore the difference between automation and augmentation in the workplace. Automation refers to AI systems taking over repetitive tasks, boosting efficiency and consistency. In contrast, augmentation means using AI to enhance human decision-making and creativity—essentially supporting rather than replacing human efforts.

AI is reshaping workforce roles and the skills employees need. Job roles are changing, with greater demand for digital literacy and critical thinking. Adaptability becomes a key asset as AI technologies continue to evolve. Organizations must invest in continuous training and reskilling programs to prepare employees for these shifts.

Implementing AI-driven automation effectively requires some best practices. Clear communication helps ensure everyone understands the purpose and impact of automation. Engaging stakeholders early builds support and alignment. Following ethical guidelines protects trust and promotes responsible AI use. And finally, continuous evaluation ensures your AI initiatives stay relevant and effective over time.

To conclude, leading transformation with AI is both a challenge and an opportunity. Change leaders have a unique role in guiding their organizations through AI-driven transformation. By embracing AI strategically, fostering digital skills, addressing ethical concerns, and supporting the workforce, leaders can ensure their organizations thrive in this new era. AI is not just a technology—it's a catalyst for meaningful business change, and with the right leadership, its potential is limitless.

Chapter 3: The Change Management Lifecycle and AI

Change management is a structured approach to transitioning individuals, teams, and organizations to a desired future state. It starts with creating awareness about the need for change — helping people understand why the change is happening. Next, we build desire and knowledge among stakeholders, ensuring they're motivated and understand what's required. Finally, we develop the ability to implement change and reinforce it, so it becomes embedded in the organization's culture.

AI plays a significant role in modern change management. First, it automates data analysis, quickly revealing trends and insights to support decision-making. AI also predicts potential outcomes, helping organizations foresee risks and opportunities. Through AI, we can personalize communication strategies, ensuring messages are relevant and engaging. Most importantly, AI delivers actionable insights, boosting an organization's ability to adapt quickly during change initiatives.

Integrating AI into change processes brings major benefits, like improved decision-making thanks to rapid data analysis and increased efficiency by automating repetitive tasks. But it also comes with challenges. Data privacy is a key concern when handling sensitive information, and organizations need skilled professionals who can manage and optimize AI tools effectively.

Traditional change readiness assessments often rely on manual surveys and data analysis, which can be time-consuming and limited in scope. AI-driven assessments, on the other hand, use advanced data analytics and machine learning to deliver faster, deeper insights. This enables organizations to better evaluate how prepared they are for change and take action accordingly.

AI tools are powerful when it comes to evaluating organizational preparedness. They can predict readiness levels, customize interventions based on specific organizational needs, and ultimately increase the chances of a successful transformation. By analyzing vast amounts of data, AI provides leaders with the insights needed to make informed decisions before launching change initiatives.

Predictive models offer a smarter way to analyze stakeholders. By using historical data, we can better understand stakeholder behaviors and trends. Incorporating behavioral insights allows us to forecast reactions more accurately, which in turn helps us create proactive strategies for managing stakeholder support — and resistance — during change.

AI helps categorize stakeholders based on factors like influence and interest, which makes it easier to prioritize engagement efforts. It can also forecast how stakeholders are likely to react, allowing organizations to shape their strategies accordingly. With these insights, you can develop targeted engagement plans that address the specific needs and concerns of different stakeholder groups.

AI-driven insights are key to enhancing stakeholder engagement. By analyzing stakeholder data, AI helps tailor communication strategies that resonate on a personal level. This personalization not only improves communication effectiveness but also increases stakeholder buy-in, reducing resistance and fostering stronger support for the change initiative.

One of the most impactful uses of AI is in personalizing communication strategies. With AI, organizations can craft messages that are tailored to specific audience segments, making them more relevant and engaging. This ensures that each stakeholder receives communication that speaks directly to their interests and concerns.

AI is transforming content creation and distribution. Tools powered by AI can quickly generate targeted content for different audiences. They also help ensure that content is distributed efficiently across multiple channels, maintaining consistency in messaging both internally and externally. This ensures that everyone receives the right message at the right time.

Measuring the effectiveness of communication is crucial — and AI excels at this. It tracks communication reach in real time, so you know who's seeing your messages. AI also performs sentiment analysis, gauging the emotional responses of your audience. Plus, it measures engagement metrics, allowing you to adjust strategies quickly and maximize impact.

AI enables continuous, real-time tracking of change adoption metrics. This helps leaders get timely insights into how well a change initiative is being adopted. By monitoring key indicators like usage rates and employee sentiment, organizations can maintain clear visibility into progress and quickly respond to any emerging issues.

When it comes to analyzing feedback and resistance, AI is a game changer. It processes both quantitative and qualitative feedback, giving a comprehensive view of stakeholder responses. AI can also detect patterns of resistance, helping organizations understand underlying challenges and guiding them to implement targeted interventions that address the root causes.

Continuous improvement is vital in change management. AI continuously analyzes data and generates recommendations that help refine strategies over time. Change leaders can use these insights to optimize their approaches and sustain momentum. With AI, the focus shifts from reactive adjustments to proactive, data-driven improvements.

To wrap up, AI is transforming every stage of the change management lifecycle — from readiness assessments and stakeholder analysis to communication planning and adoption monitoring. By leveraging AI, organizations can drive more effective, efficient, and sustainable transformations. As you move forward, consider how these AI tools and techniques can enhance your own change initiatives.

Chapter 4: AI in Organizational Diagnostics

Let's start by understanding the fundamentals of organizational diagnostics. This involves systematically assessing the health, effectiveness, and dynamics of an organization. Traditionally, this has been done through surveys, interviews, and performance data — but AI opens up entirely new possibilities for deeper, faster, and more accurate insights.

Artificial Intelligence plays several crucial roles in modern diagnostics. First, advanced data analysis: AI leverages machine learning and natural language processing to handle complex datasets that would overwhelm human analysts. Second, precision in diagnostics: AI helps uncover hidden patterns in data, improving the accuracy of insights beyond what traditional methods can offer. And third, dynamic diagnostic processes: With AI, diagnostics can adapt in real time as new data comes in, making detection and monitoring far more responsive and effective.

Chapter 4: AI in Organisational Diagnostics

AI in organizational contexts offers both exciting benefits and important challenges. On the benefits side, it boosts efficiency by automating routine analysis and offers deeper insights through advanced data processing. But challenges include data privacy concerns — organizations must handle sensitive information carefully. There's also the risk of algorithmic bias, where AI might replicate existing biases in data, potentially leading to unfair outcomes. Finally, human oversight is critical. Even with AI's power, decisions should remain accountable and ethical, guided by human judgment.

So, what is sentiment analysis and how does it work? At its core, sentiment analysis uses AI to evaluate text-based data — like surveys, emails, or social media posts — to determine the emotional tone behind words. It helps organizations measure employee sentiment on a large scale, highlighting patterns of satisfaction, concern, or disengagement that might otherwise be missed.

Why is this important? Analyzing employee feedback with sentiment analysis reveals key trends in engagement and morale. It allows organizations to spot issues early — giving them a chance to proactively address concerns, foster a positive workplace culture, and strengthen employee engagement before problems escalate.

However, best practices and ethics are essential in sentiment analysis. Transparency and consent are critical — employees need to know when and how their data is used. Bias mitigation is also important; sentiment analysis tools must be carefully monitored to avoid biased interpretations. And maintaining trust through strong data privacy practices is vital — mishandling this can seriously damage employee confidence in leadership.

Now, let's look at network analysis within organizations. Network analysis helps us map out how people really connect and communicate — beyond formal structures. It gives insight into who collaborates, shares information, and holds influence, even if they don't have an official leadership title.

One key application is identifying influencers and informal leaders. By mapping social networks, we can see who naturally holds sway within teams and across departments. These informal power holders often have more influence over workplace dynamics than formal managers.

Organizations can work with these individuals to drive communication and lead change initiatives more effectively.

When used well, network analysis can have a significant strategic impact. It helps design tailored interventions suited to the organization's unique dynamics. It also enhances collaboration by identifying and connecting key players within the network. And when organizational strategy is aligned with actual social dynamics, the chances of success increase — making strategies more realistic and grounded in reality.

Behavioral analytics is another powerful AI application in diagnostics. It involves collecting and analyzing data on employee behaviors — such as work patterns, collaboration habits, and digital interactions. This gives insight into the overall organizational climate, helps gauge engagement levels, and assesses readiness for change or transformation initiatives.

One of the most impactful uses of behavioral analytics is predicting readiness for change. AI models can analyze patterns in behavior to detect early signs of resistance or support for transformation efforts. This predictive insight allows organizations to prepare and manage change initiatives much more effectively, reducing the risk of failure.

But behavioral analytics works best when integrated with other diagnostic tools. It helps reveal patterns in behavior, while sentiment analysis captures emotional tone, and network analysis uncovers relationships. Together, they provide a comprehensive diagnostic picture — giving leaders richer insights and better guidance for action. An integrated approach ensures organizations aren't just relying on one perspective but are viewing their dynamics holistically.

To conclude, AI brings advanced capabilities to organizational diagnostics — enabling deeper insights, better predictions, and more responsive strategies. But it's crucial to balance this with ethical considerations, human oversight, and a holistic integration of diagnostic tools. By doing so, organizations can truly harness AI's power to foster healthier, more effective, and more resilient workplaces. If you need, I can also prepare a summary script for the video intro or outro.

Chapter 5: AI-Powered Change Communication

AI plays a significant role in change communication by making messaging more streamlined, personalized, and responsive. It helps automate and optimize communication workflows, ensuring clear and timely messages reach employees during organizational change. With intelligent technologies, you can create personalized outreach, tailoring communications to individual needs and preferences. AI also enables feedback analysis, helping you better understand employee concerns and improve engagement throughout change initiatives.

Traditional change communication faces several challenges. First, generic, one-size-fits-all messaging often misses the mark with diverse employee needs. Second, delayed feedback makes it harder to adapt messages quickly, which limits their effectiveness. And finally, limited opportunities for engagement can result in employees feeling disconnected and less invested in the change process.

AI-driven strategies are transforming how we approach communication. By leveraging data-driven personalization, AI can tailor messages to specific audiences based on real data insights. Automated support systems, like chatbots, ensure timely responses and reduce manual workload. And with real-time insights, organizations can adjust communication strategies on the fly, increasing their effectiveness and engagement.

Natural Language Processing, or NLP, is a key AI capability in communication. It allows AI to understand and interpret human language, making automated communication more meaningful. NLP also powers sentiment analysis, helping you gauge emotions and opinions in employee feedback. Finally, NLP can even generate tailored content, enabling personalized communication at scale.

There are several techniques for crafting personalized messages with NLP. Sentiment analysis helps you read the emotional tone of employee feedback. Dynamic text generation allows AI to create content that's customized and responsive to individual needs. And audience segmentation lets you categorize employees based on their preferences, ensuring the right message reaches the right person.

Personalized messaging brings real benefits for employee engagement. It makes messages more relevant, increasing their impact. By reducing unnecessary information, it prevents overload and helps employees focus on what really matters. And importantly, it fosters a stronger emotional connection, boosting engagement and acceptance of change.

Conversational AI, including chatbots and virtual assistants, is reshaping how organizations handle FAQs and user support. These tools use AI to deliver fast, accurate, and consistent responses, freeing up your teams for higher-value tasks.

Chatbots can handle frequently asked questions with efficiency and consistency. They help reduce the workload on HR and communication teams by taking care of routine inquiries. And most importantly, they provide timely answers to employees, which improves satisfaction and ensures your communication remains responsive.

Beyond answering questions, conversational AI enhances user support during times of organizational change. It provides ongoing assistance, addresses concerns promptly, and supports smoother transitions by maintaining open lines of communication with employees.

Measuring communication effectiveness starts with clear performance indicators. Message reach tells you how many people saw your communications. Engagement rates show how actively people interacted with your messages. Sentiment scores reflect how people felt about the content. And feedback quality measures how meaningful and useful their responses were.

AI analytics play a vital role in monitoring and optimizing communication. By analyzing responses in real time, AI helps you quickly assess engagement and sentiment. Continuous monitoring allows you to fine-tune your strategies for better results. And insights from AI empower you to make timely adjustments that maximize impact.

Using feedback and data insights, you can keep improving your communication strategies. Data-driven improvements help refine your approach for better outcomes. And by adapting strategies iteratively based on continuous feedback, you maintain relevance and increase effectiveness throughout the change process.

In conclusion, AI-powered change communication offers powerful tools for enhancing how organizations connect with employees during change. By leveraging AI for personalization, support, and measurement, you can drive higher engagement, better address concerns, and ensure your communication strategies are both effective and adaptable.

Chapter 6: Adaptive Planning and Scenario Modelling

Adaptive planning is all about adjusting strategies dynamically as new information emerges. It allows organizations to stay flexible and responsive in ever-changing environments. By enabling strategy modifications in real time, adaptive planning helps businesses navigate uncertainty, make better decisions under pressure, and maintain a competitive edge in volatile markets.

Scenario modelling lets organizations explore different possible futures to anticipate risks and prepare for various outcomes. There are two main approaches. Qualitative techniques use narratives and storylines to examine potential events and uncertainties, providing a broad perspective. Quantitative techniques rely on data-driven models and simulations, projecting likely future conditions and giving measurable forecasts. Together, these approaches offer a comprehensive toolkit for strategic planning.

AI plays a key role in making adaptive planning more effective. First, AI can analyze vast amounts of data to uncover meaningful patterns that inform strategy. It also enables real-time scenario updates, giving decision-makers the agility to respond quickly as situations change. Ultimately, AI enhances decision-making by providing insights that help leaders adapt to complex and rapidly evolving environments.

AI excels at identifying potential risks before they escalate. By processing historical data and spotting trends, AI can highlight emerging threats that might be overlooked by traditional analysis. This proactive approach allows organizations to mitigate risks more effectively and stay ahead of potential disruptions.

AI-driven tools help organizations assess risks with greater accuracy and efficiency. For instance, risk severity assessment tools analyze past incidents to predict the potential impact of future risks. Likelihood prediction models estimate the chances of various risk events happening. By integrating data from multiple sources, AI supports comprehensive risk management strategies that are proactive rather than reactive.

Let's look at some real-world examples of AI in proactive risk management. In industries like finance, manufacturing, and supply chain management, AI has been used to predict risks ranging from market shifts to equipment failures. These case studies show how AI can not only anticipate risks but also help organizations take preventive action, ultimately protecting their operations and investments.

Generative simulations use AI to create diverse scenarios that explore a wide range of possible futures. By simulating different outcomes, organizations can better prepare for uncertainty and ensure their plans remain flexible. These simulations help decision-makers visualize potential paths forward and develop strategies that are robust against change.

AI is especially powerful when it comes to simulating complex systems involving multiple variables and interactions. These simulations help organizations understand how different factors influence each other, making it easier to predict the dynamics of change. By using AI-driven models, businesses gain deeper insights into how different scenarios might play out and the risks they may face along the way.

Evaluating simulation outcomes requires a clear understanding of key metrics and their implications. It's important to manage uncertainty by recognizing its role in the simulation results and adjusting interpretations accordingly. AI enhances this process by providing advanced visualization tools and analytics, making it easier to extract actionable insights from simulation data and support better decision-making.

AI helps optimize strategic roadmaps by analyzing plans and identifying potential bottlenecks. It maps out dependencies between different tasks, helping teams understand how changes affect the broader project. AI also uncovers opportunities for efficiency improvements, providing data-driven insights that refine planning and execution.

AI-driven algorithms play a vital role in dynamically optimizing project timelines and resource allocation. They analyze project data to suggest better ways of scheduling and allocating resources, making project plans more adaptable. With real-time scheduling, teams can adjust swiftly to changes, improving both flexibility and overall project efficiency.

Continuous improvement is key to long-term success, and AI supports this through feedback loops. By constantly monitoring performance, AI identifies areas for enhancement and suggests iterative improvements. These insights ensure that planning and strategies remain aligned with organizational objectives, adapting as new outcome data becomes available.

To wrap up, AI-driven adaptive planning and scenario modelling empower organizations to navigate uncertainty with confidence. By leveraging AI for risk management, generative simulations, and strategic roadmap optimization, businesses can make more informed decisions and remain competitive in a rapidly changing world.

Chapter 7: AI for Talent and Capability Assessment

AI is becoming a key player in HR and talent management. For recruitment, AI helps by analyzing candidate data to make better hiring decisions and reduce bias in the process. When it comes to performance evaluation, AI enables more accurate assessments using data-driven insights and even real-time feedback. And for employee engagement, AI tools can personalize communication and monitor satisfaction, ensuring employees feel valued and heard.

Integrating AI into workforce capability assessments brings several benefits. First, it allows for a more objective analysis of skills and potential, minimizing human bias. It also speeds up the process of identifying high-potential talent, making recruitment and promotions more efficient. Furthermore, AI enables personalized development plans tailored to each employee's strengths and career goals. Finally, aligning workforce capabilities with company strategy becomes easier, leading to improved organizational performance.

However, it's important to consider both ethical and practical factors when applying AI. Ethically, organizations must address concerns like data privacy, transparency, fairness, and actively work to mitigate bias in AI systems. On the practical side, challenges include ensuring the quality and integrity of data being used, as well as effectively managing change within the organization during AI implementation.

AI helps analyze and identify workforce skills in several ways. By processing resumes with natural language techniques, AI can accurately extract relevant skills and experience. It also analyzes performance data to evaluate strengths and pinpoint skill gaps. This comprehensive analysis allows organizations to better understand their workforce and identify areas for growth and training.

Various technologies and platforms support AI-driven skills mapping. These platforms often integrate data from HR and learning management systems to provide a comprehensive view of skills across the organization. Dynamic dashboards make it easier for leaders to visualize skills data, and predictive analytics help forecast future skills needs and guide workforce development strategies effectively.

Let's look at some real-world examples of organizations successfully using AI for skills mapping. These case studies highlight how companies have leveraged AI tools to better understand their workforce's capabilities, close skill gaps, and align talent with business needs — often leading to measurable improvements in efficiency and strategic alignment.

AI also plays a significant role in personalizing learning and development. By analyzing learner profiles — including strengths, weaknesses, and preferences — AI can recommend customized learning paths. These tailored recommendations ensure employees acquire the right skills more efficiently, making learning initiatives both more relevant and more impactful.

Aligning training with organizational goals is critical. AI helps correlate skills data with key business objectives, identifying exactly where training is needed. With these insights, training programs can be strategically designed to directly support the company's priorities. This approach ensures that learning and development efforts contribute meaningfully to organizational success.

Measuring the effectiveness of AI-driven learning initiatives is crucial. AI tools can track learning outcomes, monitor engagement levels, and even assess business impact. This helps organizations ensure their learning programs are delivering real value — not just in terms of learner progress, but also in driving business performance and demonstrating a clear return on investment.

AI analytics support workforce planning, especially during times of change. By identifying employees' transferable skills, AI makes it easier to redeploy people across roles when needed. AI can also highlight targeted learning opportunities, ensuring upskilling efforts are aligned with business demands. This agility helps organizations adapt to shifting needs more effectively.

AI also helps mitigate risks and maintain business continuity. By predicting future skill shortages, AI enables proactive workforce planning. It also supports early risk identification, giving organizations time to manage potential impacts. And during transitions, AI helps ensure operations continue smoothly, minimizing disruption to the business.

To wrap up, AI offers powerful tools for talent and capability assessment, skills mapping, learning and development, and workforce planning. By using AI effectively and ethically, organizations can enhance their workforce strategies, make smarter decisions, and remain agile in a constantly changing business landscape.

Chapter 8: Responsible AI in Organizational Change

Let's start by defining what Responsible AI means in organizational transformation. First, ethical AI design—this means embedding ethical principles right into the AI's system design to reflect an organization's values. Then there's transparency—ensuring the AI's decisions are understandable and clear to all stakeholders. And importantly, fair and accountable decisions—this means making sure AI systems support decisions that are unbiased and that responsibility can be clearly assigned.

Why are organizations adopting responsible AI practices? Regulatory compliance is a big driver—staying on the right side of the law and avoiding penalties. Building stakeholder trust is also crucial—when people trust your AI systems, they're more likely to accept and adopt them. There's also risk mitigation—reducing both operational risks and ethical pitfalls. And finally, enhancing decision quality—responsible AI leads to decisions that are not only accurate but also fair.

Let's look at some real-world benefits and challenges. On the positive side, responsible AI can improve decision accuracy by leveraging reliable data analysis. It also increases trust and supports ethical standards, strengthening the organization's reputation. However, challenges like poor data quality can undermine these benefits. And there's often resistance to change and technical complexity—both of which can slow down or complicate AI adoption.

Transparency is critical in AI-powered decision-making. It helps stakeholders understand how AI decisions are made, which builds clarity and trust. It also reduces uncertainty, making outcomes more predictable. Transparency promotes accountability, showing exactly how decisions happen. And finally, it supports ethical participation, giving stakeholders a voice in how AI is used.

So how do we make AI more explainable? One way is model simplification—using simpler models that people can understand. Visualizations also help by showing how decisions are made through data patterns. Feature importance analysis tells us which factors most influence outcomes. And counterfactual explanations show how small changes in input could lead to different decisions, which really helps in making AI behavior clearer.

How do we ensure stakeholder trust during organizational transformation? We need open communication—keeping people informed and engaged throughout the process. Stakeholder involvement is also key—letting them have a say in AI design ensures their concerns are heard. Finally, we need transparency and inclusivity—being open about AI's role fosters confidence and promotes a sense of inclusion.

Let's move on to bias and fairness in change analytics. Bias in AI can come from skewed data, wrong assumptions, or model limitations.

48

Detecting bias means auditing data and algorithms to check if any groups are unfairly affected by AI-driven changes.

How do we measure and reduce bias? We can use fairness metrics to quantify bias and check for equitable outcomes. Bias correction algorithms help detect and correct bias in decision-making. Using diverse training data reduces bias and makes models more generalizable. And with continuous monitoring, we can catch new biases that might emerge over time.

There are plenty of real-world examples of fairness challenges in AI-driven change analytics. Bias often emerges from poor data or flawed algorithms. Ignoring these issues can lead to unfair outcomes and a loss of trust in AI systems. But we can tackle them with strategies like bias detection, data balancing, and using fairness-aware algorithms.

Ethical dilemmas are a big part of AI-driven automation. Job displacement is a real concern, with automation threatening employment in many sectors. Decision accountability is tricky too—who's responsible when AI makes a bad call? Privacy concerns arise with how AI collects and uses data. And then there are unintended consequences, which can be social or organizational impacts we didn't foresee.

It's important to balance the efficiency gains of automation with its human impact. Ethical frameworks help guide this process by setting out principles and guidelines.

So, what do these frameworks say? They emphasize transparency—making sure automated systems are clear and open to scrutiny. They promote fairness and bias prevention to ensure decisions are equitable. They demand accountability measures—so someone is always responsible for AI outcomes. And they stress respect for human rights—protecting privacy, dignity, and freedom even as automation advances.

To wrap up, responsible AI in organizational change isn't just a nice-to-have—it's essential. By focusing on transparency, fairness, and ethics, organizations can drive positive transformation while minimizing risks. This approach not only builds trust but also ensures that AI supports both business goals and societal values.

Chapter 9: Change Governance in the Age of AI

As AI becomes increasingly integrated into organizations, change governance must also evolve. AI introduces both significant opportunities and complex challenges. We'll examine how governance approaches need to adapt to address these emerging dynamics while supporting responsible AI deployment and transformation efforts.

AI presents several key challenges and opportunities for change governance. Algorithmic Bias: AI can reflect or amplify biases in data, leading to unfair decisions. This requires strong ethical oversight. Workforce Adaptation: AI adoption changes workforce needs. Organizations must plan for reskilling and address job displacement concerns. Data Privacy Concerns: The vast data AI uses raises privacy issues, making robust data governance and compliance essential.

Leadership plays a critical role in navigating AI-driven change. Leaders must champion ethical AI practices, foster a culture of transparency, and ensure their teams are ready for new ways of working. Effective leadership also involves clear communication, stakeholder alignment, and continuous learning as AI evolves.

Traditional change governance frameworks need adaptation for AI initiatives. Model Validation is crucial—ensuring AI models perform reliably in real situations. Transparency builds trust by helping stakeholders understand AI decision-making. Iterative Learning is necessary, as AI systems must continuously learn and improve over time based on new data.

Embedding AI effectively into governance processes involves key best practices: Cross-Functional Collaboration ensures diverse perspectives inform AI oversight. Continuous AI Monitoring helps identify issues early and maintain system integrity. Ethical Guidelines Embedding makes sure AI use aligns with organizational values and societal norms.

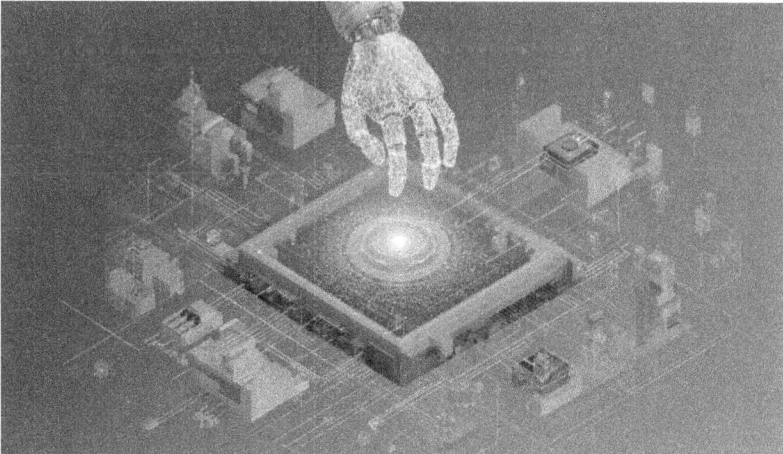

Stakeholder alignment and communication are critical for successful AI adoption: Clear Communication explains AI's goals, benefits, and limitations. Managing Expectations helps prevent misunderstandings and unrealistic goals. Addressing Concerns through open dialogue builds trust and encourages smoother adoption.

AI projects come with unique risks that governance must address: Data Quality Challenges can undermine AI effectiveness if data is poor or biased. Ethical Considerations involve risks related to bias, privacy, and fairness. Technology Obsolescence means AI systems can quickly become outdated, requiring ongoing attention and updates.

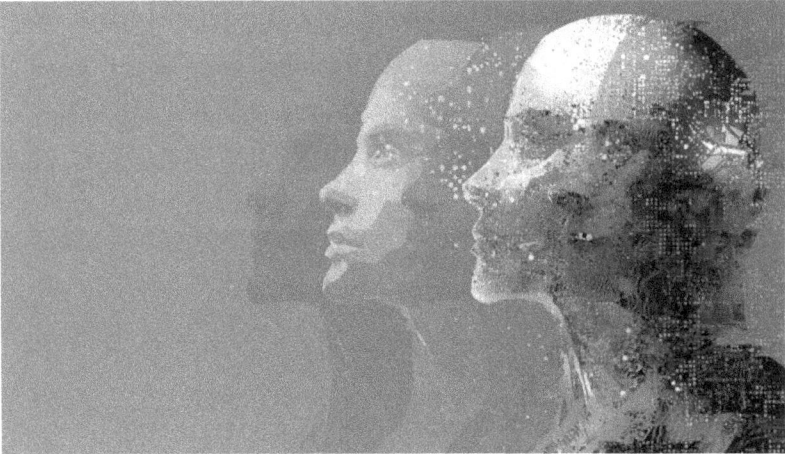

To mitigate AI-related risks, organizations should: Implement Governance Controls aligned with ethical and regulatory standards. Use Scenario Planning to anticipate potential risks and prepare appropriate responses. Conduct Continuous Risk Evaluation to identify and address risks throughout the project lifecycle.

Ensuring AI systems comply with ethical standards and regulations is vital: Adherence to Ethical Norms safeguards human rights and promotes transparency. Regulatory Compliance protects against legal risks and reinforces responsible AI usage. Building Trust through transparent practices encourages stakeholder confidence and support.

When measuring AI-driven change, key performance indicators include: Model Accuracy — how correctly the AI predicts or classifies data. Model Drift — monitoring performance changes over time. User Adoption Rates — how often the system is used in practice. Business Impact — assessing the AI's contribution to business goals and ROI.

It's important to measure AI's broader impact on organizational change: Efficiency Gains — has AI streamlined workflows or improved productivity? Employee Engagement — are employees motivated and collaborating effectively with AI tools? Decision-Making Changes — is AI enabling better, data-driven decisions within the organization?

Continuous improvement is at the heart of effective AI governance: Data Analytics delivers insights to refine AI models and ensure performance. AI System Refinement relies on learning from outcomes and making iterative improvements. Governance Practice Enhancement uses insights to evolve governance frameworks for responsible AI deployment.

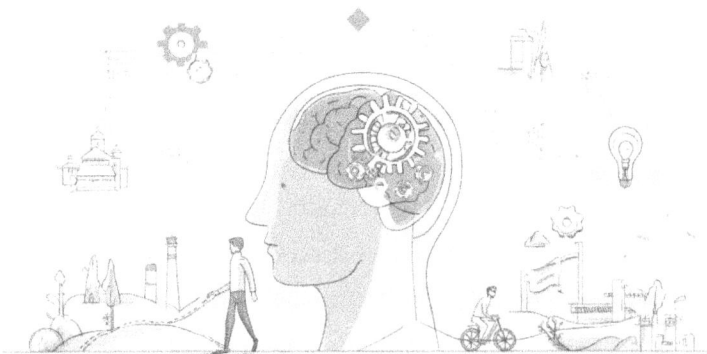

To conclude, AI-driven change governance is about balancing innovation with responsibility. By integrating ethical oversight, continuous monitoring, and clear stakeholder communication, organizations can harness AI's potential while managing its risks. Effective governance frameworks ensure that AI supports sustainable, trustworthy, and impactful organizational transformation.

Chapter 10: AI-Driven Change Management Platforms

AI-driven change management platforms are redefining how organizations handle change. By automating routine tasks, AI improves efficiency and accuracy in change processes. Predictive analysis allows companies to foresee challenges and outcomes, enabling proactive strategies. Plus, AI supports better decision-making with data-driven insights, helping organizations implement change more effectively and with less resistance.

We're seeing rapid adoption of AI platforms across industries. Companies are embracing these tools for greater efficiency and innovation. Machine learning is increasingly applied to sentiment analysis, helping businesses understand customer feedback. Communication automation tools speed up interactions and boost engagement, while predictive analytics forecast the potential impact of changes, supporting smarter, proactive decision-making.

AI addresses some of the toughest challenges in change management. It helps reduce employee resistance by enabling personalized change strategies. Communication becomes more effective as AI streamlines information sharing within teams. Perhaps most importantly, AI offers real-time insights and feedback loops, giving organizations the data they need to make informed decisions quickly during times of change.

AI-driven analytics tools give organizations deep insights into change management data, helping leaders make better decisions. Engagement tracking platforms monitor how employees are responding to change, so any issues can be addressed early. Automated workflows take care of repetitive processes, making change initiatives more efficient and ensuring a consistent approach.

Different vendors offer varying AI capabilities, so it's important to assess their strengths. Some specialize in analytics, others in automation or sentiment tracking. Integration options also differ — some platforms connect easily with existing systems, while others may require customization. Pricing models vary too, from subscriptions to enterprise licenses. And vendors often target specific markets, from small businesses to large enterprises.

When selecting a platform, look at how well it integrates with your existing systems. Check the level of AI sophistication — you want a tool that can really enhance automation and decision-making. Don't forget about user adoption support — training and onboarding resources can make or break a rollout. Finally, security and reliable vendor support are critical for long-term success.

AI-powered communication tools make it easier to share information clearly and efficiently during change processes. Predictive analytics help forecast the impact of changes, allowing for better planning. Resistance monitoring tools track employee sentiment so you can adjust strategies when needed. And adaptive learning modules ensure continuous improvement by tailoring content and training to user needs.

AI platforms should integrate smoothly with the systems your organization already uses. This seamless connection helps maintain workflow efficiency and ensures data consistency across departments. Unified data sources allow for better-informed decisions and make change initiatives more effective. Integration also ensures your collaboration and project management tools all work together to support smooth implementation.

Security is a top concern — platforms must comply with industry regulations to protect sensitive data. As your organization grows, scalability is essential, so the platform should be able to handle increased demand. Customization flexibility matters too, ensuring the platform can be tailored to your specific processes and company culture.

AI tools can significantly improve operational efficiency, saving both time and resources. They also reduce the risk of change failure by minimizing human error. By automating routine tasks, AI boosts employee engagement, freeing them up for more creative and meaningful work. Ultimately, this leads to substantial cost savings through optimized processes and smarter resource allocation.

To measure adoption success, look at user adoption rates — how many employees are actually using the new system? Employee feedback scores give insight into satisfaction and engagement levels. Tracking change initiative completion times helps assess efficiency, while measuring the impact on business objectives, like revenue growth and productivity, shows whether the change is delivering results.

Real-world examples show how organizations have successfully used AI platforms for better operations and innovation. These case studies highlight measurable benefits like increased efficiency, reduced costs, and higher customer satisfaction. They also offer valuable lessons on best practices and potential pitfalls when deploying AI-driven change management solutions.

Chapter 10: AI-Driven Change Management Platforms

To wrap up, AI-driven change management platforms offer powerful tools to streamline processes, enhance decision-making, and deliver measurable business impact. By understanding the tools, key features, integration needs, and ROI metrics, you'll be better equipped to leverage AI for successful change initiatives within your organization.

Chapter 11: Case Studies of AI in Change Management

Let's begin with digital transformation in banking. This sector has seen significant change as banks adopt AI technologies to enhance customer experiences, improve fraud detection, and optimize internal operations.

One key area is AI-driven customer service. Banks are integrating AI chatbots and virtual assistants to handle customer inquiries efficiently. These tools provide 24/7 support, reduce waiting times, and free up human staff for more complex tasks, enhancing overall customer satisfaction.

Another major impact of AI is in fraud detection. Machine learning models can analyze vast amounts of transaction data in real time, identifying suspicious patterns and activities. This means banks can detect fraud more quickly and accurately, which helps protect both the organization and its customers from financial losses.

AI is also revolutionizing banking operations through process automation. By using Robotic Process Automation and AI tools, banks can automate repetitive tasks like compliance checks and data entry. This not only improves efficiency and data accuracy but also helps reduce operational costs by minimizing the need for manual labor.

Moving on to manufacturing—AI plays a crucial role in workforce reskilling. AI analytics help organizations assess employee competencies and identify skill gaps. By analyzing both employee performance and production demands, companies can align their workforce development strategies with actual business needs.

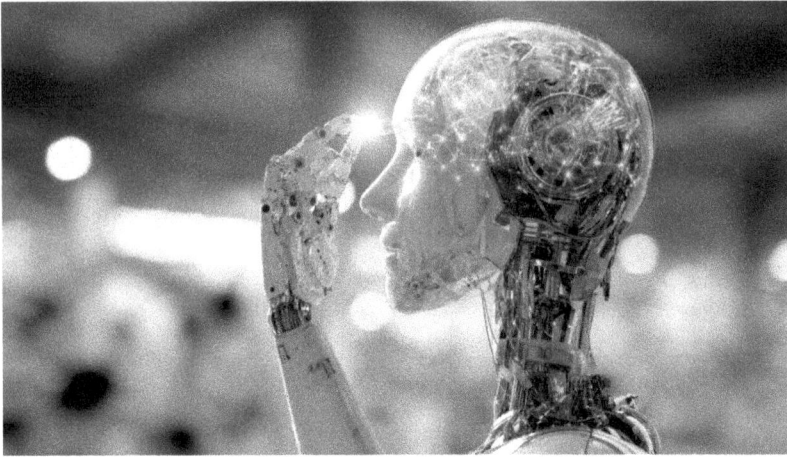

Adaptive learning platforms powered by AI personalize training for workers. These platforms tailor content to suit individual learning styles and speeds, making training more engaging and relevant. As a result, workers are more likely to retain knowledge and apply new skills effectively in their roles.

AI also supports ongoing workforce development by monitoring training progress and evaluating performance metrics. This real-time feedback helps companies measure the success of reskilling efforts and refine their strategies, ensuring continuous improvement in workforce capabilities.

Now let's look at healthcare, where predictive AI is making a big difference in patient care management. Predictive analytics models can forecast health risks based on patient data, allowing healthcare providers to intervene early and create personalized treatment plans. This proactive approach helps reduce hospital readmissions and improves patient outcomes.

AI also streamlines hospital operations through accurate forecasting. Hospitals can use AI to manage resources more efficiently, optimize staff scheduling to prevent burnout, and predict patient admission rates to maintain high-quality care even during peak periods.

Finally, AI enhances diagnostic accuracy. Machine learning models can analyze complex medical images and data, spotting patterns that might be missed by the human eye. This leads to faster diagnoses, more accurate clinical decisions, and ultimately better patient care.

To conclude, AI is reshaping change management in profound ways—whether it's enhancing customer experiences in banking, reskilling the workforce in manufacturing, or improving patient care and operations in healthcare. Understanding these applications can help organizations navigate change more effectively and harness AI's full potential.

Chapter 12: The Future of Change Leadership with AI

Let's start with an overview of change leadership in the digital era. Firstly, agility in leadership is critical. Leaders today must be agile— ready to adapt quickly to technological changes and shifting business environments. Secondly, innovation needs to become a core value. Encouraging creativity and fostering continuous improvement are key drivers of successful change initiatives. And finally, embracing new tools is essential. Leaders who adopt and leverage digital tools can better guide their organizations through transformation journeys.

Now let's discuss how AI is integrating into change management. AI enables the automation of repetitive tasks within change processes, which boosts efficiency and reduces the risk of human error. It also enhances data insights—providing leaders with deep analytics to help them better understand the impacts of change initiatives. And finally, AI tools support improved stakeholder engagement by allowing for more personalized communication and interaction throughout the change process.

With AI integration comes both opportunities and challenges for organizations. On the opportunity side, AI-driven efficiency allows tasks to be automated, enhancing predictive capabilities and overall performance. However, resistance to change is a common challenge— employees may hesitate to adapt to AI technologies and new workflows. There are also ethical concerns to consider. Organizations must use AI responsibly to maintain fairness and trust. Lastly, skill development is crucial. Companies need to invest in training employees so they're equipped to work effectively with AI systems.

Let's look at how the role of the change manager is evolving. Communication roles are shifting as AI automates routine communication tasks, freeing leaders to focus on strategic messaging and deeper engagement. Stakeholder management is also changing— automation helps streamline this process, making it easier for leaders to build relationships and make informed decisions. With many routine tasks automated, change managers can now devote more energy to strategic leadership and long-term vision.

Effective change management increasingly depends on collaboration between human leaders and AI systems. It's a partnership—combining AI's analytical strengths with human judgment leads to better decisions. By leveraging AI analytics, change managers gain powerful insights that support planning and execution. Ultimately, this collaboration enhances the success rate of change initiatives by uniting human expertise with AI capabilities.

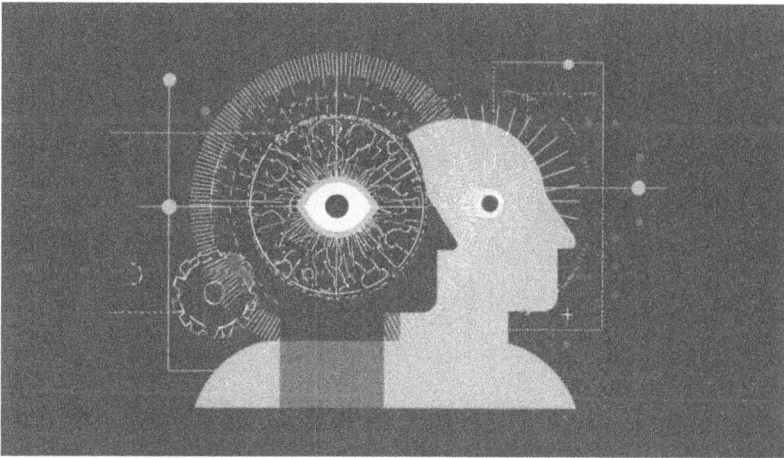

The evolving landscape also demands new skills and leadership qualities. Data literacy is now essential—leaders need to understand and interpret data effectively to make informed decisions. Technological fluency is another must-have. Being comfortable with emerging technologies is key for driving innovation and managing change. And finally, adaptive leadership is critical. Leaders need to be flexible, innovative, and resilient to navigate the complexities of AI-driven transformation.

Now, let's explore how AI is augmenting decision-making competencies. AI accelerates analysis—processing large, complex datasets quickly to deliver timely insights for decision-makers. This leads to improved decision quality, with AI providing data-backed perspectives that enhance the accuracy of strategic choices.

AI also plays a significant role in predictive analytics and scenario planning. With AI-driven predictions, leaders can anticipate future trends and outcomes more accurately. These predictive insights empower leaders to engage in effective scenario planning, allowing them to prepare strategically for various possibilities.

A critical competency is balancing human intuition with machine intelligence. AI excels in analytical tasks, delivering powerful data analysis and pattern recognition. However, human intuition and empathy bring ethical considerations and emotional intelligence into the decision-making process. By combining AI's strengths with human insight, leaders can ensure decisions are both efficient and ethically sound.

Let's talk about long-term trends and paradigm shifts in organizations. We're seeing a move toward agile and collaborative cultures— organizations are emphasizing teamwork and adaptability to foster innovation. Technology-enabled environments are becoming the norm, helping teams collaborate more effectively and streamline workflows. Additionally, flatter and more flexible organizational structures are emerging, enabling quicker decision-making and greater responsiveness to change.

Continuous learning and adaptive leadership are also key trends. Leaders must commit to lifelong learning—constantly seeking new knowledge to stay relevant in a rapidly changing environment. Adaptability is equally important—leaders need to adjust strategies and practices swiftly to keep up with technological advancements and shifting market demands.

Ethical considerations and governance are critical when implementing AI-driven change. Ethics should guide AI use, ensuring fairness and preventing bias in decision-making processes. Governance frameworks are necessary to establish accountability and transparency in AI applications. Transparency itself is vital—stakeholders need to understand how AI-driven decisions are made, which builds trust and confidence in leadership.

In conclusion, the future of change leadership with AI presents both exciting opportunities and complex challenges. Leaders who embrace AI as a partner—leveraging its analytical power while maintaining ethical standards and human empathy—will be best positioned to lead successful transformations. The key is adaptability, a commitment to continuous learning, and a focus on ethical governance, all of which will define effective leadership in the age of AI-driven change.

Lead Change with Intelligence. Navigate Transformation with Confidence.

AI in Change Management is your essential guide to understanding how artificial intelligence is redefining the way organizations plan, lead, and sustain change. As digital disruption accelerates, change professionals must rethink their approaches—integrating AI not just as a tool, but as a partner in transformation.

Whether you're a change leader, consultant, HR professional, or executive, this book equips you with the knowledge and frameworks needed to lead responsibly and effectively in an AI-driven world.

Inside, you'll discover:

- How AI is reshaping organizational diagnostics, talent analysis, and communication
- The role of AI across each phase of the change management lifecycle Practical insights into adaptive planning, scenario modeling, and risk mitigation
- Tools and platforms that are powering next-generation change strategies
- Responsible AI practices for ethical governance and inclusive transformation.

Real-world case studies that showcase AI in action across industries With a forward-looking perspective and real-world examples, AI in Change Management offers a roadmap for aligning human leadership with machine intelligence—ensuring that your organization stays adaptive, resilient, and values-driven in the face of constant change.

Part of the AI Awareness Series — practical guides designed to help professionals across sectors understand the real-world impact of AI, without the hype.